NANOTECHNOLOGY SCIENCE AND TECHNOLOGY

T0039162

SILVER NANOPARTICLES APPLIED ON PHOTONICS MATERIALS

NANOTECHNOLOGY SCIENCE AND TECHNOLOGY

Additional books in this series can be found on Nova's website under the Series tab.

NANOTECHNOLOGY SCIENCE AND TECHNOLOGY

SILVER NANOPARTICLES APPLIED ON PHOTONICS MATERIALS

YOU YI SUN

Nova Science Publishers, Inc.

New York

For permission to use material from this book please contact us:
Telephone 631-231-7269; Fax 631-231-8175
Web Site: http://www.novapublishers.com

NOTICE TO THE READER

The Publisher has taken reasonable care in the preparation of this book, but makes no expressed or implied warranty of any kind and assumes no responsibility for any errors or omissions. No liability is assumed for incidental or consequential damages in connection with or arising out of information contained in this book. The Publisher shall not be liable for any special, consequential, or exemplary damages resulting, in whole or in part, from the readers' use of, or reliance upon, this material.

Independent verification should be sought for any data, advice or recommendations contained in this book. In addition, no responsibility is assumed by the publisher for any injury and/or damage to persons or property arising from any methods, products, instructions, ideas or otherwise contained in this publication.

This publication is designed to provide accurate and authoritative information with regard to the subject matter covered herein. It is sold with the clear understanding that the Publisher is not engaged in rendering legal or any other professional services. If legal or any other expert assistance is required, the services of a competent person should be sought. FROM A DECLARATION OF PARTICIPANTS JOINTLY ADOPTED BY A COMMITTEE OF THE AMERICAN BAR ASSOCIATION AND A COMMITTEE OF PUBLISHERS.

LIBRARY OF CONGRESS CATALOGING-IN-PUBLICATION DATA

Sun, You Yi.
  Silver nanoparticles applied on photonics materials / You Yi Sun.
    p. cm.
  Includes index.
  ISBN 978-1-61728-351-2 (softcover)
  1. Photonics--Materials. 2. Nanoparticles. I. Title.
  TA1520.S86 2009
  621.36--dc22   2010022439

Published by Nova Science Publishers, Inc. ✛ New York

CONTENTS

PREFACE

Despite all the studies done so far in this field, the application of silver nanoparticles on photonics materials is a relatively new physical process. Therefore, it is quite normal that most of the work done so far has been devoted to the development and optimization of the effect rather than to a deeper understanding of the mechanism of the physical process. This book discusses mechanistic aspects of the physical process between noble metallic nanoparticles on the photonics materials, which are plausible and in line with earlier and new findings of our group, as well as the comparison of these results with those of other groups.

Chapter 1

INTRODUCTION

Recent theoretical progress in understanding the application of silver nanoparticles on the photonics materials based on the surface plasmon resonance (SPR) has been discussed in this chapter. In the first, a novel procedure to enhance the luminescence from Europium complexbased on the surface-enhanced fluorescence of silver nanoparticles, was described. It shows that the noble metal nanoparticles act as enhancer and quencher of Europium complex fluorescence. And then the both interactions strongly depend on noble metal particle diameter, concentration and surrounding medium, the systematic studies have been carried out. Secondly, recent studies about third order nonlinear optical properties of noble metal nanocomposite film were also discussed. It shows that a nonlinear optical response in copper, silver and gold nanocomposite materials with an enhanced third order nonlinear susceptibility, which is particularly useful in their applications as optical switchers with ultrashort time response and optical limiters of intense laser radiation. The crystallite size and concentration of nanoparticles on their nonlinear properties of such systems are discussed. Particularly, the synthesis method of the silver nanocomposite film is also described. Thirdly, the influence of silver nanoparticles on the phase behavior of liquid crystalline polymers was also conclued. A series of polymer films containing liquid crystalline groups and silver nanoparticles were prepared. Local electromagnetic fields near the particle can be many orders of magnitude higher than the incident fields due to hat light at the surface plasmon resonance frequency interacts strongly with metal particles and excites a collective electron motion, or plasmon. As a result, photo-induced reorientation of liquid crystalline polymers was affected by silver nanoparticles. The effect of liquid crystalline polymers structure, size and concentration of silver nanoparticles, polarized light on the photo-induced reorientation of the liquid crystalline group films was systemic studied.

Despite all the studies done so far in this field, the application of silver nanoparticles on the photonics materials here is a relative new physical process first described about 10 years ago. So it is quite usual that most of the work done so far was devoted to the development and optimization of the effect than to a deeper understanding of the mechanism of the physical process. The scope of this chapter is to discuss some mechanistic aspects of the physical process between noble metallic nanoparticles on the photonics materials, which are plausible and in line with earlier and new findings of our group, and to compare them with results of other groups. It may be a help for further discussions and the development of better optical materials based on SPR of noble nanoparticles.

Chapter 2

FUNDAMENTAL CONCEPTS

2.1. SURFACE PLASMON RESONANCE

Surface plasmon resonance (SPR) of silver nanoparticles is a surface bound electromagnetic wave propagating at the interface between free electron silver and a dielectric layer[1]. The propagation constant (k_{SP}) of the surface plasmon (SP) is dependent on the effective refractive index of the dielectric layer(n_a), which denote the sensing medium. At resonance, i.e. at a specific angle of incidence, θ_{SP} of a monochromatic beam of light, the propagation constant of the light parallel to the surface, k_x , is matched to the real part of the propagation constant of the surface plasmon, k_{SP}[2]:

$$k_x = (2\pi / \lambda)n_p \sin \theta_{SP} = \text{Re}(k_{sp})$$
$$\approx (2\pi / \lambda)\,\text{Re}[(\varepsilon_m n_a^2)^{1/2} / (\varepsilon_m + n_a^2)^{1/2}]$$
$$P = Kx \times 10^{np} \times P_0$$

where P_o and P is the intensity of input laser and surface bound electromagnetic feild, respectively; λ is the wavelength in free space; n_p is the refractive index of the prism; ε_m is the relative permittivity of the silver and Re denotes the real part of the expression. Owing to the damping of the SP caused by absorption of the light in the silver film, angles close to the resonance condition can also excite a SP. This means that a dark band, denoted the SPR dip, is projected onto a detector, as shown in Figure 1.

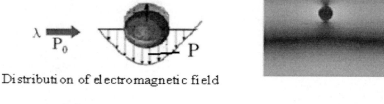

Resonance of electron cloud

Distribution of electromagnetic field

(A)　　　　　　　　　　　　　　　　　　(B)

Figure 1. (A) The SPR configuration and SPR photo graph of silver nanoparticles.

2.2. SURFACE-ENHANCED LUMINESCENCE

A procedure to enhance the fluorescence from Eu complex based on the surface-enhanced fluorescence (SEF) of noble nanoparticles has been developing[3-9]. The surface-enhanced fluorescence is attributed to the large electromagnetic field arising from the excitation of surface plasmon polariton (SPP) of silver nanoparticles. The mold of enhancement fluorescence is shown in Figure 2. According to the mold and Mie theory [10], the enhancement factor is calculated by the following equation:

$$E_1 = \left(2\frac{\varepsilon_2\varepsilon_a - \varepsilon_3\varepsilon_b}{\varepsilon_2\varepsilon_a + 2\varepsilon_3\varepsilon_b}\frac{r_2^3}{r^3}+1\right)E_0 COS(\theta)e_r + \left(\frac{\varepsilon_2\varepsilon_a - \varepsilon_3\varepsilon_b}{\varepsilon_2\varepsilon_a + 2\varepsilon_3\varepsilon_b}\frac{r_2^3}{r^3}-1\right)E_0 SIN(\theta)e_\theta$$

The E_1 is amplified incident field around silver metallic surface. The ε and P is presented in following equations:

$$\varepsilon_a = \varepsilon_1(3-2P)+2\varepsilon_2 P$$
$$\varepsilon_b = \varepsilon_2(3-P)+\varepsilon_1 P$$
$$P = 1-\left(r_1 \big/ r_2\right)^3$$

Where ε_1 and ε_2 are electrical constant of silver metal and medium around silver metal, respectively. r_1 and r_2 is radius of pure silver nanoparticles and silver metal complexes, respectively. When the θ is supposed to be 0, the enhancing factor is shown in following equation:

$$\alpha = \left| 2 \times 10^3 \cdot \frac{\varepsilon_1 - 7}{\varepsilon_1 + 14} \frac{1}{r^3} + 1 \right|^2$$

In the Drude mode, there are several assumptions as shown in following: (1) the particles are spherical; (2) the organic materials adsorbed on surface of particles are ignored; (3) the interaction between particles is not considered. It is easy to be observed from the equation (1) that the SEF strongly depend on particle size, particles shape, concentration, the distance between Lanthanon complexes and particle, and surrounding medium.

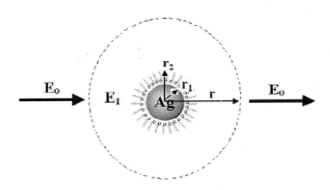

Figure 2. The representation of the amplified incident field around metallic surface and nonradiative relaxation due to damping of the dipole oscillators by the silver metallic surface.

2.3. PHOTO-INDUCED RATE

As well-known, the photo-induced rate and stabilization of LC polymers are an important parameter with regard to their potential use as the optical materials[11]. Generally, the photo-induced rate can be improved in the presence of silver nanoparticlesm, which is also attributed to the surface plasmon resonance[12]. The light of linear polalrized laser at the surface plasmon resonance frequency interacts strongly with silver particles and excites a collective electron motion, or plasmon. As a result, local electromagnetic fields near the particle can be many orders of magnitude higher than the incident fields. So, the intensity of light interacting with LC polymers can be improved, resulting in the increase of photo-induced rate. Although the mechanism has not been understood completely, total enhancement is generally believed to be a result of combination of electromagnetic and chemical effects between the adsorbed molecules and the surface. Surface plasmon resonance, which is associated with collective electron resonance induced by incident light on a rough metal surface, is similar to the SERS phenomenon[13].

2.4. THIRD-ORDER NONLINEAR OPTICAL RESPONSE

Materials exhibiting large third-order nonlinear optical response accompanied by low losses (e.g. at the wavelength of interest), good optical quality, mechanical stability and processability are generally considered as promising candidates for use in optoelectronic devices and potential photonic applications[14]. Silver has inherently very large optical nonlinearities based on the surface plasmon resonance. The Maxwell-Garnett theory has been successfully applied to the calculation of nonlinear optical properties by the following equation[15].

The silver metal core is assumed to have a displacement **D**-local electric field **E** response of the form:

$$D = \varepsilon_c(E)E = \varepsilon_c^{(0)}E + \chi_c^{(3)}|E|^2 E, \qquad (1)$$

where $\varepsilon_c^{(0)}$ is the linear dielectric function and $\chi_c^{(3)}$ is the third-order nonlinear susceptibility. Here, the second-order susceptibility $\chi_c^{(2)}$ vanishes in the present case. Thus, silver particles have nonvanishing linear dielectric function and third-order susceptibility.

Such cubic nonlinearity is the lowest-order nonlinearity appearing in the material with inversion symmetry or macroscopic isotropy. we can obtain explicit expressions for $\varepsilon_c^{(0)}$ and $\chi_c^{(3)}$

$$\varepsilon_e^{(0)} = \varepsilon_m + 3f\varepsilon_m \frac{\varepsilon_c^{(0)} + (\frac{2I}{a\varepsilon_m} - 1)\varepsilon_m}{P} \qquad (2)$$

and

$$\chi_e^{(3)} = f\chi_c^{(3)} \left(\frac{3\varepsilon_m}{P}\right)^2 \left|\frac{3\varepsilon_m}{P}\right|^2. \qquad (3)$$

It is known that the effective nonlinearity $\chi_c^{(3)}$ of the composite can be strongly enhanced by the embedding of nonlinear small silver particles in the matrix. The enhancement of $\chi_c^{(3)}$ is believed to stem from two essential elements[16]. One is the large nonlinear susceptibility of metal particles $\chi_c^{(3)}$ and the other is surface plasmon excitation. Three mechanisms, including interband transitions between the d band and the conduction band, intraband transitions in the quantum-confined conduction band, and hot electron effects, can contribute to $\chi_c^{(3)}$. Each of these mechanisms depends on the electronic description of metal particles. $\chi_c^{(3)}$ can also be enhanced due to the enhancement of the local field near the surface plasmon resonant frequency. The surface plasmon resonant frequency depends on the linear dielectric function $\varepsilon_c^{(0)}$, which is also determined by the electronic description of silver particles. From Eqs. (2) and (3), we can find, due to local field enhancement (3 ε_m/P is called linear local-field enhancement factor), both $\varepsilon_c^{(0)}$ and $\chi_c^{(3)}$ show resonance behavior. However, the surface plasmon resonant effect on $\chi_c^{(3)}$ is much pronounced. The above considerations high light n the dependence of $\varepsilon_c^{(0)}$ and $\chi_c^{(3)}$ on the electronic properties of silver particles.

It is known that, for some noble metal, the linear dielectric function of metal $\varepsilon_c^{(0)}$ can be characterized by the Drude free-electron model in combination with the Lorentz oscillator model for the bound-electron contributions. The linear dielectric function is written as

$$\varepsilon_c^{(0)} = 1 - \frac{\omega_{pf}^2}{\omega^2 + i\dfrac{\omega}{\tau_f}} + \frac{\omega_{pb}^2}{\omega_0^2 - \omega^2 - i\dfrac{\omega}{\tau_b}} \qquad (4)$$

where ω_{pf}, ω_0 are the plasmon frequency and the boundelectron resonant frequency, respectively, τ_b is the bound electron decay time, and τ_f is the free electron scattering time and is found particle-size dependent

$$\frac{1}{\tau_f} = \frac{1}{\tau_0} + \frac{V_f}{a}, \qquad (5)$$

where τ_0 is the scattering time in the bulk metal, and v_f denotes the Fermi velocity. So far, we have formulated the dependence of $\varepsilon_c^{(0)}$ and $\chi_c^{(3)}$ on the mean radius of metal particles via a simple model by introducing an interfacial factor I. The effective linear dielectric function $\varepsilon_c^{(0)}$ and cubic nonlinear optical susceptibility $\chi_c^{(3)}$ can be readily solved as a function of particle-size a by substituting Eqs.(4) And (5) into Eqs. (2) and (3).

For wavelengths near the SPR, the local electromagnetic field inside the particles is enhanced leading to a strong amplification of the third order nonlinear optical (NLO) properties of the nanocomposite as first reported by Ricard et al. According to this model, for small metal volume concentration p, the nanoconposite third order susceptibility $\chi^{(3)}$ is a function of the metal concentration, the bulk metal third order susceptibility $\chi_m^{(3)}$, the local field factor $f(\omega)$ and the incident light angular frequency p as related by the following equation:

$$\chi^{(3)} = pf(\omega)^2 \, | \, f(\omega) |^2 \, \chi_m^{(3)} .$$

For independent spherical particles, the local field factor is defined by

$$f(\omega) = \frac{3\varepsilon_d}{\varepsilon_m + 2\varepsilon_d} ,$$

where ε_m and ε_d are, respectively, the metal and matrix dielectric permittivities. It turns out that the NLO properties of nanocomposite materials are strongly dependent on the nanostructure of the films. Indeed, the metal dielectric permittivity is strongly correlated with the SPR spectral position that in turn depends on the particle size and shape. Therefore, in order to fabricate high performance NLO nanocomposite, a high number of structural parameters needs to be precisely controlled.

Chapter 3

SYNTHESIS OF OPTICAL MATERIALS CONTAINING THE SILVER NANOPARTICLES

3.1. THE SYNTHESIS OF SILVER/LANTHANON COMPLEX NANOCOMPOSITE MATERIALS

3.1.1. The Synthesis of Silver/Lanthanon Complex Composite Nanoparticles

(1) Lanthanon Ion Grafted to Silver Nanoparticles

Here, the synthesis of silver/lanthanon complex composite nanoparticles was synthesized by more than two steps as shown in Figure 3A. Firstly, the silver/functional molecule composite particles are prepared by in-situ method, in which the silver nanoparticles are formed in the presence of functional molecule. The functional molecule must contain the groups (-NH-, -S-, -OH, -COOH and so on) that can form physical or chemical interaction with silver nanoparticles and Lanthanon ion, such as poly(vinylpyrrolidone), Trifluorothenoyl-acetone, pyridine-3, 5-dicarboxylic acid and dipicolinic acid[3-5]. Secondly, the Lanthanon ion is added into the silver colloidal solution, and then the Lanthanon ion is adsorbed on surface of silver by the interaction between functional groups and lanthanon ion.

functional molecule(eg. 5-dicarboxylic acid and dipicolinic acid)

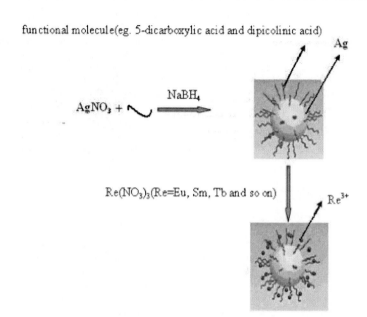

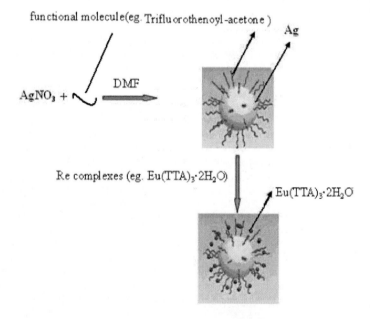

Figure 3. Synthesis of silver/lanthanon complex composite nanoparticles.

However, the fluorescence efficiency and intensity of the silver/lanthanon ion composite nanoparticles is low, which restrict their applications on optical materials. So the lanthanon complexes replacing with lanthanon ion is added into silver colloidal solution and a novel silver/lanthanon complex composite nanoparticles is formed[17], which obtain high fluorescence efficiency and intensity due the fluorescence efficiency and intensity of the lanthanon complex as shown in Figure 3B.

(2) Lanthanon Ion Grafted from Silver Nanoparticles

The silver nanoparticles was prepared in the presence of lanthanon complex as shown in Figure 4[8,17], in which the lanthanon complex contains the functional group that has strong interaction with silver. The literatures reporting the synthesis process are relatively few and only lanthanon complex $Eu(TTA)_3 \cdot 2H_2O$ as stabilizer was reported. However, the quenching fluorescence was observed due to the distance between lanthanon ion and silver metallic surface to be close zero. A novel macromolecule ligand-Lanthanon complex was used as stabilizer, in which the fluorescence quenching was restricted due the long chain of polymer and the process has been studying by our group.

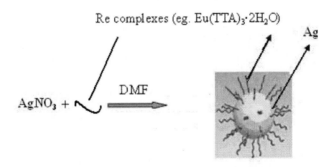

Figure 4. Synthesis of silver/lanthanon complex composite nanoparticles.

3.1.2. The Synthesis Silver/Lanthanon Complex Composite Nanoparticles Doped Solid Materials

Generally, the silver/Lanthanon complex composite nanoparticles doped inoriganic solid materials were prepared by sol-gel route. In one case[9], the synthesis of $10B_2O_3\text{-}90SiO_2$ solid materials were prepared by the sol-gel route

You Yi Sun

as shown in Figure 5. Firstly, the nanometer-sized Ag colloids was synthesized by reducing AgNO$_3$ with KBH$_4$ in the aqeuous solution containing PVP (or other funcational polymer). Secondly, the Ag collodial solution was coperated into 10B$_2$O$_3$-90SiO$_2$ solid materials together with Lanthanon ion. At first, TMOS was added dropwise to distilled water that included a small amount of HCl, and stirred. TMB in methanol (C$_3$H$_9$BO$_3$/CH$_3$OH) was added dropwise with further stirring. The polymer-protected Ag sol was mixed with methanol dissolving EuCl$_3$, and added to the B$_2$O$_3$-SiO$_2$ sol. And then, distilled water including NH$_3$ was added with stirring. It took several days to obtain a stiff gel.

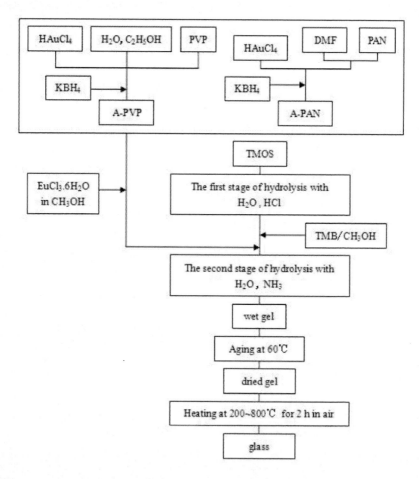

Figure 5. Synthesis of silver/lanthanon complex doped solid materials.

The composites of silver/Lanthanon complex composite nanoparticles doped polymer were also prepared either through melting extrusion and solution cast or through polymerization of lanthanide–monomer complexes. But there are few works reporting the synthesis process and the systmic study is required. As above discussions, during the synthesis process of silver/Lanthanon complex composite nanoparticles doped solid materials, the good dispersion of the complex nanoparticles on the solid matrix is a challenge and very important to obtain enhancement fluorescence.

3.2. THE SYNTHESIS OF LC POLYMER/METAL NANOCOMPOSITES

Lots of attentions were paid to some LC polymer systsiems that are usually used as templates for the preparation and stabilization of inorganic nanoparticles[18-19]. However, the role of nanoparticles in the mesophase formation, as well as their influence on liquid crystal properties, was yet to be understood. Especially, the synthesis and properties of LC polymer/silver nanopartics nanocomposites, there are only three literatures reporting the physical process as shown in following.

3.2.1. The in-Situ Method

The silver/polymer LC complex composite nanoparticles or complex film was synthesized by more than two steps as shown in Figure 6[20-21]. Firstly, the LC functional polymer is prepared, in which the functional molecule must contain the functional group (-NH-, -S-, -OH, -COOH, CN and so on) that form physical or chemical interaction with silver nanoparticles, such as poly(6-[4-(4-cyanophenylazo)phenoxy]x-methylene methacrylate), polyimide and so on. Secondly, the silver/polymer LC complex composite nanoparticles are prepared by in-situ method, in which the silver nanoparticles are formed in the presence of functional molecule. Generally, the LC polymer dose not dissolved in aqueous solution. So, the reaction is carried out in DMF or DMF/water mixing solution, in which the NaBH$_4$ or DMF solution acts as the reducer of Ag ion. In addtion, in order to obtain the silver/polymer complex composite film, the concentration of LC polymer is high during the synthesis process. And then, the nanocomposite film containing LC polymer can be formed by the spin-coating technique.

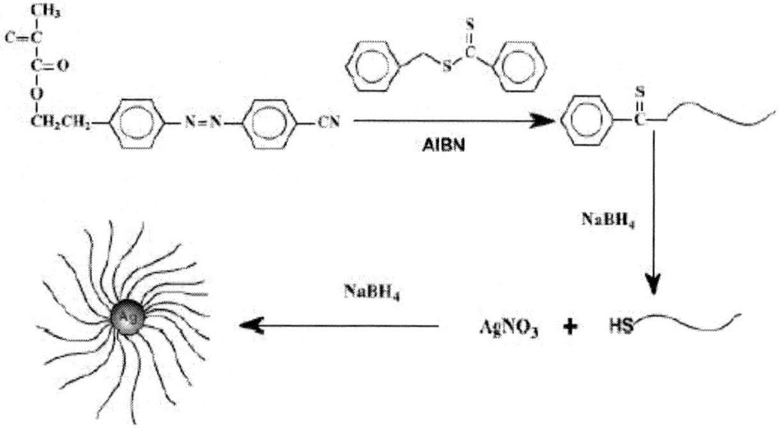

Figure 6. Synthesis of silver/LC polymer complex film.

The excellence of the synthesis method is that the silver nanoparticles can been well dispersed in LC polymer matrix and there are few aggregate of silver nanoparticles. Furthermore, the group of polymer can be located close to surface of silver nanoparticles. There are very important to enhance the photo-induced rate of LC polymer based on the SPR of silver nanoparticles. However. There are two disadvantages. Firstly, the concentration of silver doped in LC polymer is difficult to be controlled. Secondly, if the LC group has strong interaction with silver metal, the photo-induced rate will be reduced.

3.2.2. The Mixing Method

Here, the silver/polymer composite nanoparticles were also synthesized by mixing method[12]. Firstly, the silver colloidal solution is prepared by in-situ method. Secondly, the LC polymer is added into the silver colloidal solution, and then the mixing solution can also form film by the spin-coating technique. Here, the concentration and size of silver doped in LC polymer is easy to be controlled. At the same time, the strong interaction between LC group and silver metal is averted, which is benefit to enhance the photo-induced rate.

However, the poor dispersion of silver nanoparticles in LC polymer matrix decrease photo-induced rate.

3.3. THE SYNTHESIS OF THIRD-ORDER NONLINEAR OPTICAL MATERIALS

Although there are lots of works reporting the synthesis of noble metal doping solid materials[22-24], the effect of noble metal nanoparticles on third order nonlinear optical properties of these nancomposites is few consideration. The reason is that the high dispersion of sliver nanoparticles in solid materials is very important for study of third order nonlinear optical properties, which is difficult to be obtained.

3.3.1. Synthesis of Ag Nanocomposite Polymer Films

(1) The in-Situ Method
Geneally, silver metal nanocomposite polymer film was prepared by in-siut synthesis technique as shown in Figure 7, in which the silver nanoparticles was well dispersed in polymer matrix[25]. The synthesis techniques is required with a functional group for the polymer, which can be interacted with noble metal, i.e. cyano (-CN) groups, thiol (-SH) group, and so on.

However, most of polymers do not contain the functional group, for example poly(methyl methacrylate) (PMMA). As a polymer matrix, the PMMA has attracted particular interests for its low optical absorption, refractive index tailorability with molecular weight, simple synthesis and its low cost. These characteristics make it suitable as a host material for investigation and application in optical materials.

The functional sulfur end group of PMMA was introduced by Reversible addition-fragmentation transfer (RAFT) polymerization. In following, Ag nanocomposite PMMA film can be prepared by in-situ synthesis technique as shown in Figure 7. Compared to conventional synthesis procedure, it will be better for meeting two advantages: (1) it is suit to lots of monomers, which can be polymerized by RAFT to produce a polymer with such a functional group at the chain end of the polymer; (2) molecular weight and its distribution of the polymer can be controlled at the same time, providing an opportunity to investigate the effects of molecular weight and its distribution on nonlinear optical properties of nanocomposite polymer film;(3) the difference is that part of the PMMA is adsorbed on the surface of Ag nanoparticles with one end group of polymer as soon as Ag nanoparticles formed.

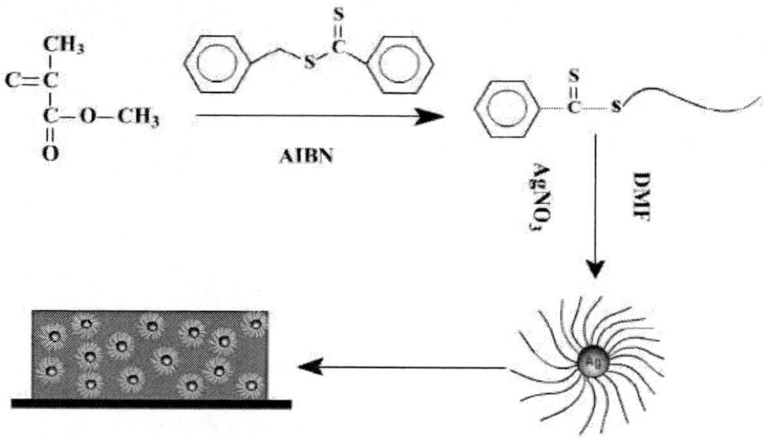

Figure 7. Synthesis of silver nanoparticles doped composite film by the in-siut method.

(2) Self-Assembly Method

Noble metal nanocomposite polymer film was also prepared by the self-assembly method, by which the high dispersion polymer film was obtained as shown in Figure 8[26]. In one case, PAH/PSS/Ag bilayer film was prepared by the method as shown in following. Firstly, the AgNO₃ solution and tannic acid solution were added to water while stirring continuously. The resulting colloid was yellow-brownish and could be stable for several weeks. Secondly, single crystal Si(111) and glass slide were used as the substrates, which were cleaned with a concentrated sulfuric acid/hydrogen peroxide solution. After rinsing with plenty of water, the cleaned substrates were then placed into a dilute aqueous solution of PEI and held. Then rinsing with water and flushing with N_2 flow, a thin layer of PEI was formed on the substrate. Prior to the construction of the Ag/PAH multilayered coating systems, two-cycle PAH/PSS bilayers were deposited on the PEI modified substrate to promote the combinability between the surface and the first monolayer of Ag film. The two-cycle PAH/PSS bilayer film architecture was achieved by alternately depositing PSS and PAH from their aqueous solutions.

As above discussions, the synthesis is very complex, which is difficult to be controlled. At the same time, few polymers as matrix are suit to the synthesis process. Addtionally, the mechnical properties of polymer film is bad, which is difficult to be applied on optical materials.

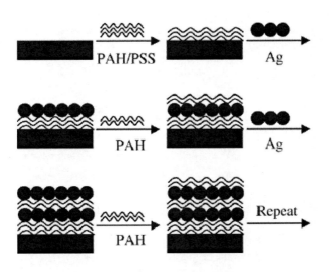

Figure 8. Synthesis of silver nanoparticles doped composite film by the self-assembly method.

(3) Others

As well-known, the good dispersion of silver in polymer matrix is very important to study the third order nonlinear properties of nancomposite. There is one work reporting the synthesis of nancomposite with high dispersion as shown in Figure 9[27]. Moreover, the present method shows an easy processing and the synthesis process is easily controlled. Aqueous solutions of silver nitrate and PVA were mixed and stirred; the solution mixture was spin-coated either directly on quartz or on glass substrates previously coated with a PS layer. Silver nanoparticles were generated by heating the solid films in a hot air oven under ambient atmosphere. PVA acts simultaneously as the reducing agent, stabilizer for the nanoparticles, and the matrix for homogeneous distribution and immobilization. But, the third order nonlinear properties of the composite film prepared by the method are not understood. There are requested for lots of works studying synthesis and third order nonlinear properties of the polymer nanocomposite in the future.

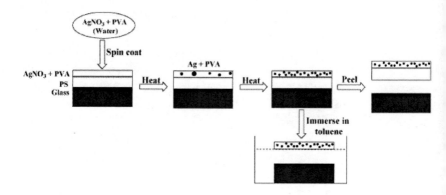

Figure 9. Synthesis of silver nanoparticles doped composite film by the new method.

3.3.2. Synthesis of Ag Nanocomposite Inorganic Films

(1) Multitarget Magnetron Sputtering Method

Some Ag nanocomposite inoriganic films (eg. Ag:Bi$_2$O$_3$ composite films) were prepared by co-sputtering of Ag and BiO onto the fused silica substrate and using the multi-target magnetron sputtering system[28]. As shown in Figure 10, the Ag and BiO targets were diagonally mounted on two target holders with a distance separation between them and at an inclined angle related to the surface normal of the substrate. The Ag and BiO target was sputtered by the DC and RF. power supply, respectively. The sputtering power on the target was controlled to achieve a variation of the Ag concentration in the film over a wider range. In the film depositing process, the sample holder was rotated to improve the uniformity of composition distribution of the film. As above discussions, the synthesis is very complex. Especially, it needs special equipment.

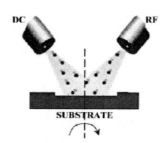

Figure 10 The Scheme of RF-Dc Co-sputtering system.

(2) Hybrid Deposition Technique

The Ag nanocomposite inoriganic films were also prepared by a hybrid process simultaneously combining PECVD and pulsed DC sputtering as shown in Figure 11[29]. The substrates (c-Si and fused silica) were placed on an RF-powered substrate holder electrode on which negative substrate bias voltage V_B develops. The working gas flow consisted of Ar, O_2, and SiH_4, while the total pressure was maintain. Ag was simultaneously sputtered from a gold target installed on a magnetron head located at a 15cm distance from the substrate. The Ag concentration was controlled by the total power delivered to the target from a pulsed DC power supply using a pulse frequency.

(3) Ion Implantation

The Ag nanocomposite inoriganic films were also prepared by ion implantation as shown in Figure 12[30-31]. The silicate glasses (SG) were prepared as plates with the thickness of 1mm. The accelerate energies used for implantation were 30 keV-60 keV at a dose. The concentration of silver and thickness of the composite film is controlled by the accelerate energies, the dose and implantation times. It is very convenience to synthesis Ag doped inoriganic films. However, the distribution and size in matrix is difficult to be controlled. In the other way, it needs special equipment.

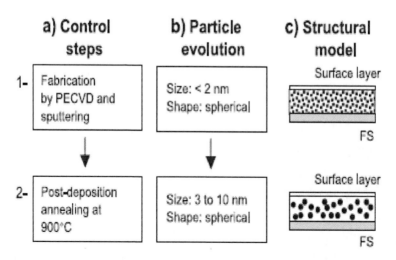

Figure 11. Illustration of the film preparation steps, evolution of the particle size and shape, and corresponding microstructural models.

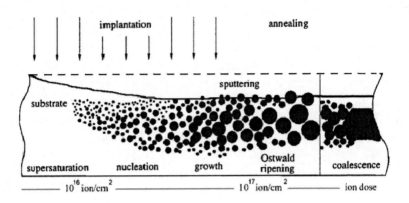

Figure 12. Scheme of major physical stages under metal ion implantation with dose increase.

Chapter 4

CHARACTERIZATIONS OF
OPTICAL MATERIALS

4.1. FLUORESCENCE PROPERTIES

As well-known, the fluorescence intensity, efficiency and lifetime of luminescent materials are very important with regard to their potential use as the optical materials [32], which are studied and analysis by RF-5301PC spectrofluorometer and lifetime spectrofluormeter (Fluorolog-3-TAU), respectively.

(1) Fluorescence Intensity

The fluorescence intensity can be obtained according to the emission spectrum as shown in Figure 13, in which the value of longitudinal coordinates show the fluorescence intensity. In one case, five emission peaks centered at 580, 592, 616, 650, and 698 nm, assigned to $^5D_0 \rightarrow {}^7F_0$, $^5D_0 \rightarrow {}^7F_1$, $^5D_0 \rightarrow {}^7F_2$, $^5D_0 \rightarrow {}^7F_3$, and $^5D_0 \rightarrow {}^7F_4$ transition of Eu ion, respectively, were observed[33]. The result indicate that the effect of silver colloids on the energy level of Eu ion is slight. However, it can see the increase in intensity by the presence of Ag colloids, while the emission wavelength remains unchanged. Of course, the fluorescence intensity can be compared under identical instrumental conditions.

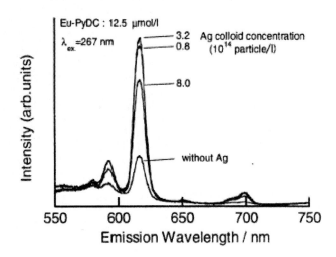

Figure 13. Emission speactrum of Eu ion complexe with various silver colloidal solution.

(2) Quantum Yields and Fluorescence Lifetime

The quantum yields (qx) have been determined by absolute measurements(RF-5301PC spectrofluorometer) and is thus determined as follows[34]:

$$q_x = \left\{\frac{1-R_{st}}{1-R_x}\right\}\left\{\frac{\Delta\varphi_{st}}{\Delta\varphi_s}\right\}q_{st} \tag{1}$$

where Rst and Rx are the amount exciting radiation reflected by the standard and by the sample, respectively, and qst is the quantum yield of the standard phosphor. The terms $\Delta\varphi$st and $\Delta\varphi$s give the integrated photo flux (photons s^{-1}) for the sample and the standard phosphors, respectively. Generally, the stand was sodium salicylate, which has a broad emission band with the maximum at 616 nm and $q = 60\%$ at room temperature [35].

The values of $R_{ST,}$ $R_{s,}$ $\Delta\Phi$,and $\Delta\Phi_{ST}$ must be obtained for the same excitation wavelength, geometry and instrumental conditions.Unmost care was taken to ensure a constant and reproducible position for the sample/standard holder and unchanged instrumental conditions throughout the measurement.The samples and standards were thoroughly ground to a fine powder into an agate mortar in order to minimize grain size difference.All measurements were carried out using a compacted powder layer to prevent insufficient absorption and back scattering of the exciting radiation.

The values of $\Delta\Phi$,and $\Delta\Phi_{ST}$ are determined by integrating the emission intensity over the total spectral range in the emission spectra plotted as quanta per wavelength interval (photons $s^{-1}nm^{-1}$) versus wavelength(mm). The emission spectra must be previously corrected for the spectral dependencies of the photomultiplier response and the grating reflectivity.

The reflection coeffocients are established by scanning the emission monochromator through the excitation wavelength region and integrating the intensities of the spectra thus obtained.In order to have absolute values.

The fluorescence lifetime can be obtained accroding to the fluorescence decay curve of Eu complex solution as shown in Figure 14. The experiment data satisfied very well the monoexponential Eq. (2)[36]:

$$I(t) = I_1 \exp(-t / t_1),$$

where I(t) was the fluorescence intensity varying with t and t_1 is lifetime.

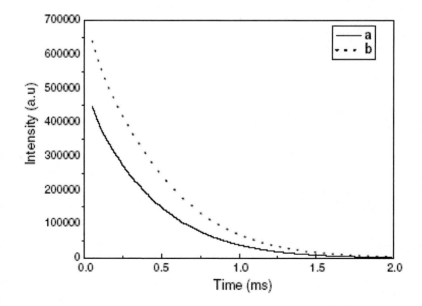

Figure 14. Fluorescence decay curve of Eu ion complexe.

4.2. PHOTO-INDUCED MOLECULAR REORIENTATION

The LC polymer can be oriented in solid matrices by polarized light due to the accompanying process of trans–cis–trans isomerization [37]. The trans-LC polymer are stable with an elongated molecular form and the cis-LC polymer are photo-induced isomers with a bent form and revert back to trans form thermally or by light.

The experiment of photo-induced anisotropy is performed by use of polarized exciting beams of two colors to investigate the features of optimization of reorientation of the LC polymer molecules. Figure15 shows the experimental arrangement. The sample is placed between two crossed polarizers (vertical P and horizontal P0) and a weak He–Ne 633 nm beam is used to probe the photo-anisotropy of the sample. Initially no light reaches the detector due to the random distribution of the azobenzene molecules. When an exciting beam polarized at 45° to the vertical from a 442nm He–Cd laser, 533nm Nd-YAG or 488nm Ar+ irradiates the sample, the analyzer transmits some 633nm light. This photoinduced anisotropy is due to the reorientation of LC polymer molecules induced by the pupming light. The LC chian transition moment lies along the molecular axis and only the molecules with their orientation parallel to the electric vector absorb light. The repeated trans–cis–trans isomerization results in the alignment of LC polymer molecules in the direction perpendicular to the polarization of the exciting beam.

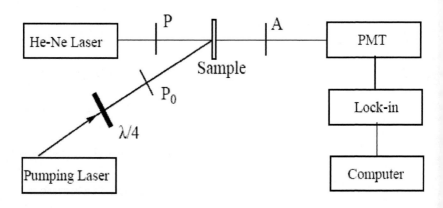

Figure15. Schematic of experimental arrangement for optimizing photo-induced molecular reorientation by using two polarized actinic beams of two colors.

4.3. THIRD-ORDER NONLINEAR
OPTICAL PROPERTIES

The third-order nonlinear optical properties have been investigated using the Z-scan technique in the nanosecond regime[26, 38]. The experimental setup used in this work is schematically shown in Figure 16. Z-scan measurements were performed using a Q-switched Nd-YAG laser emitting at 442nm, 533nm, 488nm or 1064 nm. Laser pulses were converted in second harmonic pulses resulting in the final 532 nm wavelength. A low wavelength pass filter was used to remove any remaining power. Generally, the sample measurements were performed using a 10Hz repetition rate in order to minimize the nonlinear thermal effect and sample damaging.

The laser beam was spatially-filtered and focused by a 100mm focal lens to obtain a beam radius at the beam waist. The beam profile was measured using a CCD camera. The laser beam was separated into a reference arm and a sample measurement arm using a silica wedge. A wedge was used instead of a beam splitter in other to avoid interferences in the focal volume. The intensities of the reference and signal beams were measured using standard photodiodes linked to a digital oscilloscope with a reduced 20MHz bandwidth. Special attention was taken to set-up reference and signal arms as similar as possible. Thus, optical paths of the reference line and the sample line were kept close to each other in order to minimize the measurement delay between pulses. Furthermore, incoming powers were set at similar values in order to obtain as identical photodiode responses as possible. Finally, the sample was mounted on a motorized translation stage.

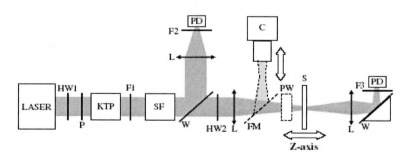

Figure 16. Experimental setup for P-scan and Z-scan measurements. Laser. HW1: half-wave plate (1064 nm). P: polarizer, KTP frequency doubling crystal. F1: absorbing filter for 1064 nm, SF: spatial filter, W: wedge, L: lenses, F2 and F3: attenuators, FM: mirror on a flip mount, C: camera mounted on translation axis, HW2: half-wave plate (532 nm), PW: pyrometer, S: sample mounted on a translation stage, PD: photodiode.

Figure 17 presents the results of Z-scan experiment for the nanocomposite film[25]. the third-order susceptibility can be considered as a complex value

$$\chi^{(3)} = \text{Re}\chi^{(3)} + i \text{ Im}\chi^{(3)}$$

Where the real part ($\text{Re}\chi^{(3)}$) is connected with the nonlinear refraction and the imaginary part ($\text{Im}\chi^{(3)}$) is connected with the nonlinear absorption as shown follow equation(1) and (2). The nonlinear refraction and nonlinear absorption of pure PMMA was negligible:

$$\text{Re}\chi^{(3)} = n_2 \cdot n_0 / 3\pi \tag{1}$$

$$\text{Im}\chi^{(3)} = \beta \cdot n_0 \cdot \varepsilon_0 \cdot c^2 / \omega \tag{2}$$

For the calculations of nonlinear refractive indices (n_2) and nonlinear absorption(β) of the samples we used the equations (3) and (4),respectively. Where ω is the laser radiation frequency and ε_0 is the electric permittivity of free space.

$$n_2 = \Delta T_{p-v} / [0.406 \, (1 - S)^{0.25} \cdot |(2\pi/\lambda) \, I_0 \, L_{\text{eff}}|] \tag{3}$$

$$E(\text{я}) = (\beta Ш(\text{я}) \, Д_{\text{уаа}})^{-1} \, \text{дт}(1 + \beta Ш(\text{я}) \, Д_{\text{уаа}}) \tag{4}$$

where ΔT_{p-v} is the peak-to-valley transmission difference in the Fig.4A, $T(z)$ is dependence in open-aperture scheme as shown in Fig.4B, I_0 is the radiation intensity at the focal point, S is the transmission of the aperture and Leff is the sample's effective length.

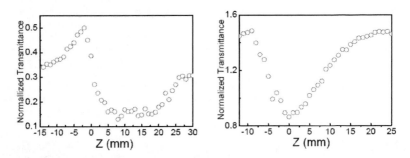

Figure 17. The dependences of the normalized transmittance in the close-aperture and open-aperture.

Chapter 5

THE EFFECTS OF SILVER NANOPARTICLES ON OPTICAL MATERIALS

5.1. THE EFFECT OF SILVER NANOPARTICLES' CONCENTRATION

5.1.1. Fluorescence

The enhancement fluorescence and the quenching fluorescence all depend on the concentration of silver nanoparticles[3-5]. Firstly, silver colloids have one of surface-enhanced phenomena based on the large electromagnetic field arising from the excitation of surface plasmon polariton (SPP). Furthermore, the enhancement field around silver metallic nanostructures is strongly enhanced when two or more particles come into close proximity with each other as shown in Figure 18 [39]. When silver nanoparticles is irradiated by light in its range of resonant frequency, the resonance plasmon of particle occurrs and make local electromagnetic fields near the particle many orders of magnitude higher than the incident fields of single particle. So the enhancement factor increases with increasing in concentration of silver nanoparticle. On the other way, the quenching effect of silver nanoparticle also increases with increasing in concentration of nanoparticles, resulting from the re-absorption of surface plasmon resonance (SPR) and silver nanoparticles scattering. So, generally the enhancement factor firstly increases with increasing in silver colloidal concentration and decreases rapidly with further increasing in concentration.

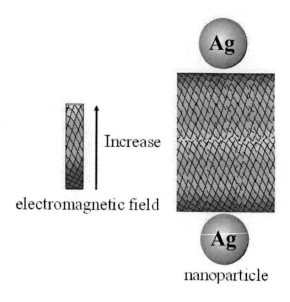

Figure 18. The representation of the large electromagnetic field between silver nanoparticles.

5.1.2. Photo-Induced Re-Orientation Rate

As well-known, silver colloids have one of surface-enhanced phenomena based on the large electromagnetic field arising from the excitation of surface plasmon polariton (SPP). And, surface plasmon polariton (SPP) around LC polymer is strongly enhanced when two or more silver nanoparticles come into close proximity with each other. On the other hand, exciting the plasmon resonance of the particles would lead to light absorption and local heating, which clearly makes it easier for the molecules to re-orientate. The above two aspects both promote the photo-induced reorientation of LC groups especially for the fast growth progress that was thought to depend on the quantum yield of the isomerization reaction, the isomerization rate and the local mobility of the LC polymer molecule. So, it is clearly seen that the re-orientation rate rapidly increases with the increase of the silver content as shown in Figure 19. On the other way, however as the sliver content increased further, the enhancement effect began to reduce. The result is attributed that the silver as filler on the polymer matrix reduce the free volume of polymer chian, which effective inhibit the movement of LC of polymer[12]. Eventually, the reorientation rate of the doped sample was even lower than that of the undoped sample. So, generally the reorientation rate at least for "fast" progress can be largely enhanced when using an appropriate dopant content of silver nanoparticles.

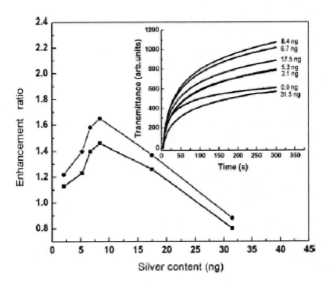

Figure 19. The re-orientation rate of LC polymer depended on the concentration of silver nanoparticles.

5.1.3. Third-Order Nonliner Optical Property

The third-order nonliner optical property is also strongly depended on the silver content [25, 28]. As well-known, the the intensity of large electromagnetic field increases with increase in silver content. So, generally, the third-order nonlinear susceptibility incerases with with increase in silver content, whereas neither dissolved in normal solvents nor doped in polymer film. But there should be one assumption that the particle is well dispersion in solid mterials and the interaction between particles is not considered.

YY. Sun has reported the preparation and nonlinear optical properties of Ag doped in PMMA film[25]. The result shows that the third-order nonlinear susceptibility increases with the increasing in the concentration of Ag doped in PMMA film. When the concentration of Ag doped in PMMA is 2.4%, the third-order nonlinear susceptibility is 6.22×10^{-9} esu, which close to the largest value (10^{-7} esu) of noble metallic nanoparticles embedded in inorganic materials. Here it can point out important effects for the enhancement third-order nonlinear susceptibility due to good distribution.

At the same reason, there are particle coalescence or percolation, the third-order nonlinear susceptibility will decrease. In one case, the nonlinear optical properties of Ag:Bi_2O_3 films is also measured for with different Ag concentration are plotted in Figure 20[28]. It can be seen that the value of third-order nonlinear susceptibility is enhanced with increasing Ag concentration up to 35.7%, and then rapidly decreases. There are two factors determining the enhancement of effective nonlinearity for metal-dielectric composites: the nanoparticle density [40] and the local filed near and inside particle [41]. For Ag:Bi_2O_3 films with relatively low Ag concentration (<35.7%), the increase of Ag concentration mainly results in an increase of particle density and decrease of interparticle spacing that enhances the multiple particle electromagnetic interaction as above-mentioned discussion about the SPR absorption, leading to an enhancement of the local field. Both effects can enhance the effective third-order nonlinear susceptibility. At higher Ag concentration, due to particle coalescence or percolation, the local field will be averaged out [42], leading to the significant weakening of the enhancement effect.

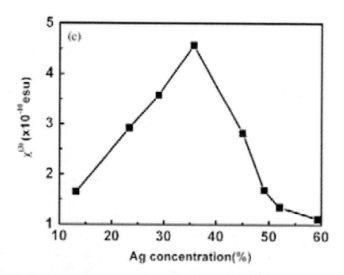

Figure 20. The third-order nonliner optical property of LC polymer depended on the concentration of silver nanoparticles.

5.2. The Effect of Nanoparticles' Size

5.2.1. Fluorescence

The enhancement and the quenching fluorescence are all mainly attributed to the SPR absorption of silver particles [3-5]. And the SPR absorption of silver nanoparticles strongly depends on their size [43]. It is obviously seen from the Table 1 that the enhancement factor decreases with increasing in size and size distribution of silver nanoparticles[3-7]. The obtained maximum enhancement factor for each sample is thus regarded as the result of a delicate balance between the enhancing and the quenching effect. The SPR absorption peak is red shift with increasing in size of silver nanoparticles. So, the resonant energy level (350nm) between ligand and Lanthanon ion is far from absorbing region of Ag colloids and the intensity of located electromagnetic field decreases with increasing in size of particles, resulting in a decrease of enhancement factor.

Since the kind of small silver colloid shows characteristic absorption band in Uv-visible region, the decrease of enhancement factor is also due to re-absorption of surface plasmon resonance (SPR) of silver colloids as shown in Table 1. The absorption bands of particles are 320nm-500nm, 320nm-700nm and 320nm-800nm corresponding to the size of 9.7nm, 15.7nm and 27.1nm, respectively. The emission photon of Lanthanon complexes (wavelength of 615nm) is re-absorption and scattered by the particles with size of 15.7nm and 27.1nm. Contrarily, the quenching effect due to re-absorption of surface plasmon resonance is almost ignoring at the size of 9.7nm. Therefore, SPR re-absorption of Ag colloids must be taken into consideration in order to improve the enhancement factor.

And each sample shows a different particle concentration at which the enhancement factor reaches the maximum (C_{max}) as shown in Table 1. The enhancement C_{max} is strongly determined by the size of the silver nanoparticles, too. The decreasing tendency in C_{max} values with increase in the particle size further indicates the increase in the quenching effect arising from the two interactions(re-absorption and scattering).

In summary, the observed fluorescence intensity is thus regarded as the result of the delicate balance between enhancing and quenching effect of the silver nanoparticles, which are all strongly dependent of the particle size. Generally, smaller particles had larger enhancement factor and the higher particle concentration region (C_{max}). Finally, it is pointed out that the observed changes in the fluorescence properties are attributed to the formation of the particle aggregates in the solution. It should also be noted that the enhancement fluorescence was observed despite the off-resonant condition between the excitation wavelength (276 nm) and the surface plasmon (ca.400nm), and these findings suggest that the improvement of enhancement factor is expected even in the solution phase by optimizing the design of both nanoparticles and Lanthanon complexes.

Table 1. The effect of the particle size on the enhancement factor and C_{max}

Size(nm)	Absorption peak (absorption band)	Enhancement factor	C_{max}
9.7	400nm(320nm-500nm)	>1.35	>1.5 ×10^{-9} mol/L
15.9	420nm(320nm-700nm)	1.35	6.0×10^{-10} mol/L
27.1	460nm(320nm-800nm)	1.15	8.0×10^{-11}mol/L

5.2.2. Photo-Induced Re-Orientation Rate

The enhancement of reorientation rate also depends mainly on size of silver particles. But, there are few works reporting the effect of silver nanoparticles' size on the reorientation rate.

For approving the above mechanism, the polymer films containing LC groups and silver nanoparticles were prepared[12]. Photo-induced reorientation was performed on the films under polarized light with wavelength at 442nm, 532nm and 365nm, respectively. The influence of silver nanoparticles on the photo-induced reorientation of the films was studied, in which the SPR bands is at 450nm. It is obviously seen that the largest enhancement was obtained when film samples were irradiated with 442nm light that is identical to the maximum absorbance wavelength of silver nanoparticles at 450nm wavelength. The enhancement was weaker when irradiating with light at 532nm because a weaker absorption can be detected at about 532nm wavelength in sliver nanoparticles' absorption spectrum. Moreover, the enhancement disappeared when induced by 365nm wavelength light for no plasmon resonance absorption of Ag nanoparticles at 365nm wavelength. The above results provide a strong evidence for resonance plasmon effect of silver nanoparticles on reorientation enhancement of LC groups.

At the same mechanism, when the wavelength of pumping laser is 442nm, the SPR bands are closer to the wavelength and the larger enhancement is obtained, in which the SPR band is controlled by the size and surface property of silver nanopartilces.

5.2.3. Third-Order Nonlinear Optical Properties

There are few works and mechanism reporting the effect of silver nanoparticles' size on the third-order nonlinear susceptibility. For approving the mechanism that the enhancement nonlinear optical absorption also deepens on their size, the nonlinear optical absorption in silver nanosol was discussed at selected wavelengths using open aperture Z-scan technique[44]. The SPR band of doped noble nanoparticles is about 550nm. The third-order nonlinear susceptibility at different wavelengths around the plasmon peak is presented in Figure 21. The NLO parameters of metallic nanoparticles embedded in an inert dielectric matrix system can be predicted using the Maxwell±Garnett theory. This theory predicts strong enhancement of the NLO properties near the surface plasmon resonance. Considering the wavelength variation of the surface plasmon enhancement factor as the most important term in governing the optical properties, a simple estimate gives a scaling of third-order nonlinear susceptibility with the square of the absorption coefficient.

The above results also provide a strong evidence for resonance plasmon effect of silver nanoparticles on third-order nonlinear susceptibility, which is also generally adjusted by the size and surface property of silver nanoparticles.

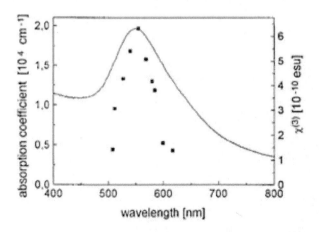

Figure 21. The third-order nonliner optical property of composite film depended on the wavelength of laser.

5.3. The Effect of Solid Phase

5.3.1. Fluorescence

The use of silver nanostructures in solid phase as fluorescence enhancers is an attractive challenge and has been studied [9, 45-51]. Enhancement fluorescence from Lanthanon ion owing to surface plasma oscillation of silver particles in solid sphase, such as $10B_2O_3$-$90SiO_2$, TeO_2-PbO-GeO_2, SiO_2 and TeO_2-PbO-GeO_2 glass were easily observed. Even if the Lanthanon ion and bare silver nanoparticles were distributed in matrix of SiO_2 glass, in which the protecting molecule or functional molecule was not used. As above discussion in solution phase, the quenching fluorescence was observed in the absence of protecting molecule or functional molecule due to Lanthanon ion being far from silver nanoparticles. Here, the Lanthanon ion can be close proximity to the silver metallic surface with the help of the solid phase. So, here it can point out important effects for the enhancement fluorescence of Lanthanon ions due to surface plasma oscillation of silver particles in solid sphase : (1) the particle sizes and their distribution, (2) the volume fraction, and (3) energy matching between the surface plasmon frequency of a particle and the excitation energy of Lanthanon ions.

In some cases, a sol-gel method was also used to prepare GeO_2-Eu_2O_3-Ag films, in which the luminescence efficiency of Eu ions during UV excitation was comparable to that in films activated by organic Lanthanon complexes. It was found to be complex Eu-Ag centers with a high quantum yield of the intracenter transfer of excitations to the rare-earth activator from silver ions and Ag oligomer clusters located on the surface of silver nanoparticles.

Additionally, enhancement and quenching fluorescence from samarium (Sm) ions owing to surface plasma oscillation of silver particles in "Aerosil" silica glasses was also investigated. The introduction of silver into the Sm-containing silica glasses prepared by the original sol-gel method leads to the formation of complex optical centers involving Sm ions and simple and/or complex silver ions. The formation of Sm-Ag centers was accompanied by an increase in the concentration of non-bridging oxygen ions, which prevent the reduction of silver ions by hydrogen. Silver nanoparticles formed in small amounts upon this reduction were effective luminescence quenchers from the corresponding excited states of Sm ions. However, the SEF of Sm ion in the solution system is few studies. So, systematic studies are required to elaborate the relation between SEF and SPR in solution with respect to the condition of silver nanoparticles for further application.

5.3.2. Photo-Induced Re-Orientation Rate

The enhancement of re-orientation rate stronghly depends on the structure of matrix LC polymer. When the LC group of the polymer has strong interaction with the silver nanoparticles, the movement of LC chain is difficult, resulting in decrease of reorientation rate. Eventually, the reorientation rate of the doped sample was even lower than that of the undoped sample. So the LC polymers structure is very important to obtain the enhancement of reorientation rate [12, 20-21]. In one case, the different structure of molecules (P1 and P2) is shown in Figure 22, in which LC chain of P1 has stronghl interaction with silver metallic spheroids. So, the formation of nanoparticles in nematic polymer matrix P1 leads to a rapid increase of the glass transition and a less-significant decrease of the clearing temperature as shown in Table 2. As a result, the overall mesophase stability is rapidly decreasing.

It shows that nanocomposites P2-Ag are able to form an LC phase, in which the movement of LC chain is not affected by the silver nanoparticles due to no interaction between P2 and silver nanoparticles.

Thses indicate that, as in P1-Ag, silver nanoparticles influence negatively the mesophase stability. However, due to the different chemical structure, P2 is less sensitive to this influence.

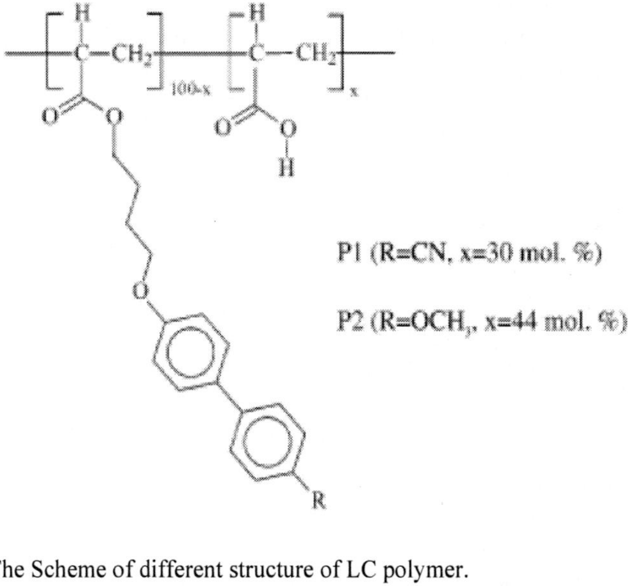

P1 (R=CN, x=30 mol. %)

P2 (R=OCH₃, x=44 mol. %)

Figure 22. The Scheme of different structure of LC polymer.

**Table 2. The glass transition and clearing temperature of
LC polymer containing silver nanoparticles**

	concentration	melt		glass	
		orign	after	orign	after
P1	2%	98	90	35	58
P2	2%	135	130	54	54
P3	50%	168	168	126	126

The data on the phase behavior and structure of the nanocomposites P1-Ag and P2-Ag discussed above let us make the first important conclusion. Nanoparticles are not passive elements of the mesomorphic polymer matrix. On the contrary, they are able to influence significantly the temperature interval of LC phase. It is of evident interest to understand the nature and mechanisms of such influence, as well as its interconnection with the chemical structure of polymer matrix. So, in oder to obtain the enhancement of reorientation rate funtions, the polymer LC should contain the funcational group that has strongh interaction with silver nanoparticles, and at the same time, the LC group has no interaction with silver nanoparticles. Generally, the polymer LC is copolymer as shown in Figure 21, and the LC chian can not contian the funtional group -CN, $-NH_2$ and so on. In the other way, the unit of copolymer contains the functional group.

However, there are lots of LC polymer, in which the LC group contain the -CN, $-NO_2$, $-NH_2$ and so on that have strong interaction with silver nanoparticles. Here, the enhancement of re-orientation rate is difficult to be observed due to the interaction inhibiting the movement of LC chain. Taking this into consideration, here a new synthetic approach to mesomorphous nanocomposite polymer system has been developed[20]. The key is that -S-groups were introduced to terminus of LC polymer by RAFT polymerization as shown in following Figure 23. It is expected to efficiently avoid the interaction between LC units and Ag because the thiol groups tend to form preferentially complexes comparing with other groups (-CN, $-NH_2$ and so on) of LC units on Ag nanoparticles[52]. As shown in Table 2, the glass and metal temperature of P3 is same with that of the P3 adsorped on silver nanoparticles. Which effectively indicate the conclusion. It will provide an opportunity to obtian enhancement of reorientation rate.

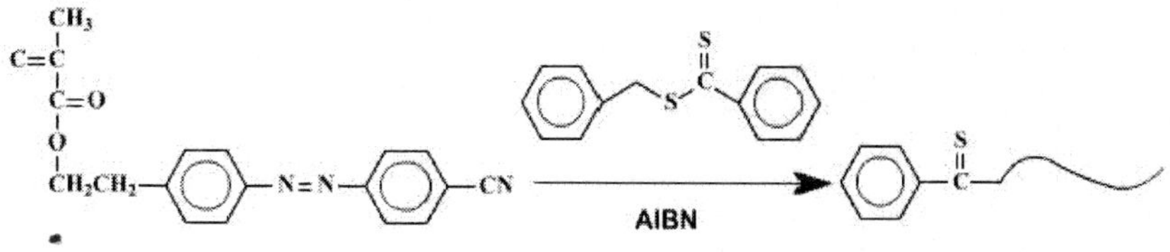

Figure 23. Schematic representation of P3 prepared by RAFT.

5.3.3. Third-Order Nonlinear Optical Properties

Let us consider a nonlinear random composite in which nonlinear spherical metal particles with concentric coating shell are randomly embedded in a linear matrix[53]. The radii of the core and the shell are a and $a+t$ (t is the thickness of interfacial layer). Dielectric constants of the linear shell and dielectric matrix are given as ε_s and ε_m, respectively. It is known that the nonlinearity $\chi_e^{(3)}$ of the composite can strongly depended on the embedding nonlinear small metal particles and the matrix as follows:

$$\chi_e^{(3)} = f \chi_c^{(3)} \left(\frac{3\varepsilon_m}{p} \right)^2 \left| \frac{3\varepsilon_m}{p} \right|^2 .$$

where $\chi_c^{(3)}$ is the third-order nonlinear susceptibility of nonlinear small metal particles ; ε_m is the Dielectric constants of the linear dielectric matrix ; f is the volume fraction of metal particles.

$$P \equiv \varepsilon_c^{(0)}(1-f) + [2(1+I/a\varepsilon_m) - f(2I/a\varepsilon_m - 1)]\varepsilon_m$$

where where $\varepsilon_c^{(0)}$ is the linear dielectric function. when I=0 (corresponding to sharp and smooth interface).

Accrding to the above Equations, The $\chi_e^{(3)}$ is determined on the $\varepsilon_e^{(0)}$ and $\chi_e^{(3)}$ of silver particles, and the dielectric properties. When the dielectric constants of the linear dielectric matrix is larger, the he $\chi_e^{(3)}$ is larger. So, for a nonlinear random composite, the property of matrix also affact the third-order nonlinear optical properties.

In some case[25], when the silver nanoparticles is doped into polymer film with the concentration of 2.4%, the composite film shows high third-order nonlinear susceptibility of 6.22×10^{-9} esu. However, the third-order nonlinear susceptibility of the silver collodial solution was measured to 1.22×10^{-11} esu, in which the Ag particles and its content were same with nancomposite polymer film. The difference is attributed to the various matrix for the silver nanoparticles, in which the dielectric constants (ca. 2.8) of the linear dielectric PMMA matrix is lower than that of DMF solution(ca.70)

Additional[19], the ultrafast nonlinear optical properties of co-sputtered silver/bismuth oxide (eg. $Ag:Bi_2O_3$) nanocomposite films with different Ag concentration (13.2–59.3 at%) were also investigated by femtosecond (fs) pump–probe and fs optical Kerr effect techniques. The result of the femtosecond OKE measurements showed that the third-order susceptibilities of $Ag:Bi_2O_3$ films have a maximum of 4.1×10^{-10} esu at Ag concentration of 35.7%.

Furthermore, nonlinear optical response of silver nanoparticles synthesized by ion implantation in silicate glasses ($\varepsilon_m=1.6$) was investigated in ultraviolet range 354.7 nm and laser radiation 1064 nm[30-31]. It was shown that $\chi^{(3)}$ in ultraviolet range (354.7 nm) of Sg :Ag was calculated to be 6.1×10^{-8} esu. It is considered that the Sg :Ag composite materials as effective homogeneous media. The legitimacy of this approach proves to be true by the fact that the sizes of MN are much smaller than the wavelength used in experiment. For an effective homogeneous medium described by the presence of resonant transitions one can apply the standard two-level model. In addtion to this, The variations of the sign of the nonlinear refractive in Nd :YAG laser radiation ($\lambda = 1064$ nm) indices depending on matrix properties are analyzed. The SG:Ag and SLSG:Ag was calculated to be 1.5×10^{-8} esu and $-3.5 \times 10-8$ esu, respectively.

Here it can point out important effects for the enhancement nonlinear optical response due to surface plasma oscillation of noble particles in glass: (1) the particle sizes and their distribution, (2) the volume fraction, (3) the particle' distribution in solid matrix, and (4) the dielectric constants of the linear dielectric matrix.

Chapter 6

APPLICATION

6.1. OPTICAL LIMITING BEHAVIOR BASED ON THIRD-ORDER NONLINEAR OPTICAL PROPERTIES

The advent of high power lasers, in both civilian and military applications, makes the search for efficient optical limiters for a wide range of personnel and equipment (e.g., sensor) protection the highest priority[54]. An ideal optical limiter exhibits a high, linear transmission below a certain "limiting" threshold, but its intensity is greatly attenuated (opaque with constant, low, and nonlinear transmittance) above the threshold. Among many promising optical limiting materials, organometallic, metallophthalocyanines, metalloporphrins, and metal clusters containing a small number of metal atoms have attracted considerable attention. Compared to organic optical limiting materials, these compounds have the advantage of multiple electronic transitions such as metal-ligand charge transfers. However, these systems often suffer from low damage threshold and inefficient optical limiting.

So, the nonlinear optical properties of silver nanoparticles are currently being explored with great interest, which is expected to suffer from high damage threshold and inefficient optical limiting. It shows that in either solid film or in solution, structurally well-defined Ag nanoparticles can achieve nearly 2 orders of magnitude of attenuation of highintensity laser power. More importantly, it opens the door to a new class of highly promising optical limiting materials based on nanosized metal clusters. In one case[55], the silver colloidal solutions were prepared by in-situ synthesis technique in the presence of the PMMA, which was polymerized by Reversible addition-fragmentation transfer. The average size of sphere silver nanoparticles is about

10nm. It shows the optical limiting properties at 532 nm as shown in Figure 24. The energy transmittance decreases when the input energy increases. The limiting threshold was about $161J/cm^2$. The limiting characteristics result from the RSA effect of the sample, which is one of nonlinear absorption mechanisms. Optical limiters have many applications, including eye and sensor protection[56], optical information processing [57] and optical communications [58].

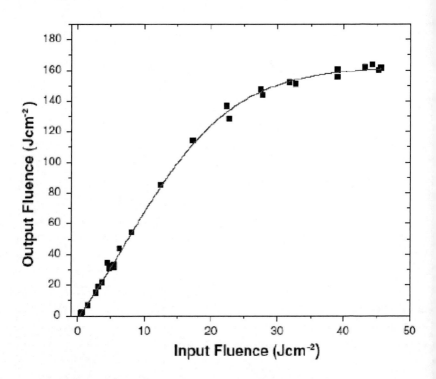

Figure 24. Optical limiting curve of the silver colloidal solution at 532nm.

In conclusion, the Ag nanomaterials in stabilized suspensions are potent optical limiters toward nanosecond laser pulses in the green, and the optical limiting properties are stable and insensitive to changes in the nanoparticle-related parameters. The optical limiting responses in the nanomaterial are likely dominated by a nonlinear absorption mechanism, though the details remain to be explored in further investigations. In addition to this, the works are in progress to design and fabricate new optical limiters based on nanosized metal clusters and to understand their optical limiting mechanism.

5.2. OPTICAL SWITCHING BASED ON THIRD-ORDER NONLINEAR OPTICAL PROPERTIES

Recent years have witnessed dramatic progress in the design of all-optical switching devices for ultrafast high band width optical communication and computing. Optical switching which enables routing of optical data signals regardless of data rate and data protocol has the advantage on conventional electrical switches to redirect the information without the expensive and power consuming optical–electrical–optical conversion. Thus, optical switching allows treatment of future data rate upgrades [59-61]. Applications of optical switching include protection and restoration in optical networks, bandwidth provisioning, wavelength routing and network monitoring.

Current interest has focused on molecular devices that offer advantages of small size and weight, extremely low propagation delay, high intrinsic speed and ability to tailor properties to suit the specific device configurations.

Recently, all-optical switching based on nonlinear excited-state absorption has been demonstrated in different molecular configurations such as liquid crystals, rubidium vapor, organometallic phthalocyanines, polydiacetylene, poly(methyl methacrylate), PVK and azobenzene dyes, polymethine dye, 2-(20-hydroxyphenyl) benzoxazole, fullerene C60 and naturally occurring retinal protein bacteriorhodopsin[62-64].Organometallic compounds are prospective nonlinear materials, as they exhibit large optical nonlinearities. Organometallic compounds have a number of advantages over organic compounds. But, these systems often suffer from low intrinsic speed and ability to tailor properties. So, the silver nanoparticles are currently being explored with great interest for application on optical switching based on third-order nonlinear optical effects, which is expected to suffer from high intrinsic speed and ability to tailor properties.

In one case[25], The $\chi^{(3)}$ of the Ag nanoparticles doped PMMA film was obtained by Z-Scan technique and increased with increase in concentration of Ag doped in PMMA. When the concentration of Ag doped in PMMA is 2.4%, the $\chi^{(3)}$ is 6.22×10^{-9} esu.

The possibilities of materials' applications as optical switching elements are further analyzed. The critical parameter that applies for estimation of optical switching effectiveness is the ratio between nonlinear absorption coefficient and nonlinear refractive index at the investigated wavelength[65].

$K = \beta\lambda/n_2$

Where n_2 is the nonlinear refractive index in SI units. In the case of Ag nanocomposite materials, the K is 8.48, 0.26 and 1.25 corresponding to concentration of 0.3%, 0.8% and 2.4%, respectively. The materials at concentration of 0.8% lead to $K< 1$, and therefore makes the material the prospective ones for optical switching applications at the wavelength of 532 nm.

It is expected that the homo-size Ag nanopaticles can be obtained and controlled if amphiphilic copolymer with surfer end group is further introduced in the technique, and corresponding investigations on synthesis and optical properties are currently in progress.

Chapter 7

CONCLUSIONS

Surface-enhanced optical properties based on the surface plasmon resonance (SPR) of silver metal nanoparticles has been dicusssed on improving fluorescence properties of Lanthanon complexes, nonlinear optical properties of composite and reorientation rate of LC polymer. At the same time, the observed quenching fluorescence is attributed to the re-absorption of SPR and photon scattering by silver metal nanoparticles, interaction between optical materials and silver nanoparticles. And then the enhancing and quenching effect strongly depend on silver metal particle size and concentration.There are lots of problemes to be investigated in the future. In some cases, the effect of distance between Lanthanon complexes and silver nanoparticles on SEF has few considerations, resulting in being relatively low enhancement factor (<10). Furthermore, it has few reports on optical memory polymer materials, nonlinear optical polymers materials and polymer optical fiber perform based on the surface-enhanced fluorescence and enhancement of photo-induced re-orientation rate. In addition, the effec of silver particles size and shape on mechanism and physical process of third-order nonlinear optical properties and enhancement of photo-induced re-orientation rate have been fewly reporting. The chapter discussions are very value to further improve overall enhancement factor and C_{max}, which are key to be applied in optical materials for the surface-enhanced optical properties phenomenon of silver nanoparticles.

REFERENCES

[1]   Knut Johansen, Ralph Stålberg, Ingemar Lundstr ¨om, Bo Liedberg, *Meas. Sci. Technol.* 11 (2000) 1630.

[2]   Kretschmann E, *Z. Phys.* 241(1971)313.

[3]   Yuehui Wang, Ji Zhou, Ting Wang, *Materials Letters,* 62(2008)1937.

[4]   H. Nabika, S. Dekia, *Eur. Phys. J. D.* 24(2003)369.

[5]   Yuehui Wang, Xinran Zhou, Ting Wang, Ji Zhou, *Materials Letters,* 62(2008) 3582.

[6]   *J. Phys. Chem.* B, 107(2003) 9162.

[7]   Youyi Sun, Zhi Zheng, Qing Yan, Jiangang Gao, Hongfang Jiu, Qijin Zhang, *Materials Letters,* 60(2006)2756.

[8]   Youyi Sun, Hongfang Jiu, Douguo Zhang, Jiangang Gao, Bin Guo, Qijin Zhang, *Chemical Physics Letters,* 410(2005)204.

[9]   Tomokatsu Hayakawa, Kazunori Furuhashi, Masayuki Nogami, *J. Phys. Chem. B.* 108(2004)11301.

[10] Wang, D. -S.; Kerk er, M, *Physical review B.* 24(1981)1777.

[11] K. Tawa, K. Kiyohara, K. Kamada, K. Ohta, Z. Sekkat, S. Kawata, *Macromolecules.* 34 (2001) 8232.

[12] Jingli Zhou, Jianjun Yang, Youyi Sun , Douguo Zhang, Jing Shen,Qijin Zhang, Keyi Wang, *Thin Solid Films.* 515 (2007) 7242.

[13] Mengtao Sun, Lixin Xia, Maodu Chen. Spectrochimica Acta Part A: *Molecular and Biomolecular Spectroscopy.* 74(2009)509.

[14] Guang Yang, Dongyi Guan, Weitian Wang, Weidong Wu, Zhenghao Chen, *Optical Materials.* 25(2004)439.

[15] R.G. Barrera, G. Monsivais, W.L. Mochan, M. Del Castillo, Physica A: *Statistical Mechanics and its Applications.* 157(1989) 369.

[16] K P Unnikrishnan, V P N Nampooril, V Ramakrishnan, M Umadevi, C P G Vallabhan, *J. Phys. D: Appl. Phys.* 36 (2003) 1242.

[17] Si Wu, Youyi Sun, Xin Wang, Wenxuan Wu, Xiujie Tian, Qing Yan, Yanhua Luo, Qijin Zhang, *Journal of Photochemistry and Photobiology A: Chemistry*. 191(2007)97.

[18] Bouchama, F.; Thathagar, M. B.; Rothenberg, G.; Turkenburg, D. H.; Eiser, E. *Langmuir*. 20(2004)477.

[19] Andersson, M.; Alfredsson, V.; Kjellin, P.; Palmqvist, A. E. C. *Nano Lett.* 2(2002)1403.

[20] Evgenii B. Barmatov, Dmitry A. Pebalk, Marina V. Barmatova, *Langmuir*. 20(2004) 10868.

[21] Jiangang Gao, Youyi Sun, Jingli Zhou, Zhi Zheng, Hongwei Chen, Wei Su, Qijin Zhang, Journal of Polymer Science Part A: *Polymer Chemistry*. 45(2007)5380.

[22] Sensors and Actuators B: *Chemical*. 117(2006)367.

[23] Jin-Woong Kim, Jung-Eun Lee, Su-Jin Kim, Jong-Suk Lee, Jee-Hyun Ryu, Junoh Kim, Sang-Hoon Han, Ih-Seop Chang, Kyung-Do Suh, *Polymer*. 45(2004)4741.

[24] A. Heilmann, C. Hamann, G. Kampfrath, V. Hopfe, *Vacuum*. 41(1990)1472.

[25] Youyi Sun, Yaqing Liu, Guizhe zhao, Xing Zhou, Qijin Zhang, Yan Deng, *Materials Chemistry and Physics*. 111(2008)301.

[26] Fredrick W. Vance, Robert C. Johnson, Buford I. Lemon, Joseph T. Hupp, Daniel L. Feldheim, *J. Am. Chem. Soc.* 122(2000)12029.

[27] Shatabdi Porel, Shashi Singh, S. Sree Harsha, D. Narayana Rao, T. P. Radhakrishnan, *Chem. Mater.* 17(2005) 9.

[28] Guanjun Youa, Peng Zhou, Chunfeng Zhanga, Zhiwei Donga, Liangyao Chen, Shixiong Qian, *Journal of Luminescence*. 119–120 (2006) 370.

[29] Jean-Michel Lamarre, Franck Billard, Chahineze Harkati Kerboua, Michel Lequime, Sjoerd Roorda, Ludvik Martinu, *Optics Communications*. 281 (2008) 331.

[30] R. A. Ganeevl, A. I. Ryasnyansky, A. L. Stepanov, T. Usmanov, *phys. stat. sol.* (b). 238(2003)5.

[31] R. A. Ganeev, A. I. Ryasnyansky, A. L. Stepanov, T. Usmanov, *phys. stat. sol.* (b). 41(2004) 935.

[32] Hongfang Jiu, Youyi Sun et.al, *Journal of Non-Crystalline Solids*. 352(2006)197.

[33] Youyi Sun, Qijin Zhang et.al, Spectrochimica Acta Part A: *Molecular and Bimolecular Spectroscopy*. 64(2006)977.

[34] O.L.Malta, H.F.Brito, J.F.S.Menezes,F.R.Goncalves esilva,C.de Mello Donega, S.Alves Jr, *Chemical Physicas Letters.* 282(1998)233.

[35] A. Bril, A.W. De Jager-Veenis, *J. Electrochem. Soc.* 123 (1976), 396.

[36] Ren-Jie Zhanga, Kong-Zhang Yanga,*, An-Chi Yub, Xin-Sheng Zhao, *Thin Solid Films.* 363 (2000) 275.

[37] Pengfei Wu , D.V.G.L.N. Rao, *Optical Materials.* 21(2002) 295.

[38] Minjoung Kyoung, Minyung Lee, *Optics Communications.* 171(1999)145.

[39] Chen, Y. L.; Fang, Y, Spectrochimica Acta Part A: *Molecular and Biomolecular Spectroscopy.* 69(2008)733.

[40] N. Pincon, B. Palpant, D. Prot, E. Charron, S. Debrus, *Eur. Phys. J. D.* 19 (2002) 395.

[41] H.R. Ma, P. Sheng, G.K.L. Wong, *Topics Appl. Phys.* 82(2002) 41.

[42] H.B. Liao, R.F. Xiao, H. Wang, K.S. Wong, G.K.L. Wong, *Appl. Phys. Lett.* 72 (1998) 1817.

[43] Senoy Thomas, Saritha K Nair, E Muhammad Abdul Jamal, S H Al-Harthi, Manoj Raama Varma, M R Anantharaman. *Nanotechnology.* 19(2008)1.

[44] Georg Maxein, Harald Keller, Bruce M. Novak, Rudolf Zentel, *Adv. Mater.* 3(1998)338.

[45] Hayakawa, T. Selvan, S. T.; Nogami, M, *Journal of Non-Crystalline Solids.* 259(1999)16.

[46] Almeida, R.d.; da Silva, Davinson M.; Kassab, Luciana R.P.; de Arau' jo, Cid B, *Optics Communications.* 281(2008) 108.

[47] Petit, L.; Griffin, J.; Carlie, N.; Jubera, V.; García, M.; Hernández, F.E.; Richardson, K. *Materials Letters.* 61(2007)2879.

[48] Geddes, Chris D.; Cao, Haishi I.; Gryczynski.; Gryczynski, Z.; Fang, J.Y.; Lakowicz, J. R, *J. Phys. Chem. A* 107(2003)3443.

[49] Malashkevich, G. E.; Shevchenko, G. P.; Serezhkina, S. V.; Pershukevich, P. P.; Semkova, G. I. ; Glushonok, G. K, *Physics of the solid state.* 49(2007)1891.

[50] Malashkevich, G. E.; Semchenko, A. V.; Sukhodola, A. A.; Stupak, A. P. ; Sukhodolov, A. V.; Plyushch, B. V.; Sidski, V. V.; Denisenko, G. A. *Physics of the Solid State.* 50(2008)1464.

[51] J.A. Jime'nez, S. Lysenko, H. Liu, *Journal of Luminescence.* 128 (2008) 831.

[52] Evgenii,B.P.;Dmitry,A.B.;Marina,V, *Langmuir.* 20(2004)10868.

[53] Jean-Michel Lamarre, Franck Billard, Chahineze Harkati Kerboua, Michel Lequime, Sjoerd Roorda, Ludvik Martinu, *Optics Communications.* 281(2008)331.

[54] Tutt, L. W.; Boggess, T. F. Prog. *Quantum Electron.* 17(1993)299.

[55] Yan Deng, Youyi Sun, Pei Wang, Douguo Zhang, Xiaojin Jiao, Hai Ming, Qijing Zhang, Yang Jiao, Xiaoquan Sun, *Current Applied Physics.* 8(2008)13.

[56] S. Shi, W. Ji, S.H. Tang, J.P. Lang, X.Q. Xin, *J. Am. Chem. Soc.* 116(1994) 3615.

[57] S.W. Koch, W. Peyghambarian, H.M. Gibbs, *J. Appl. Phys.* 63(1988)1.

[58] D.M. Russ, Proc. SPIE 630(1986) 161.

[59] G.I. Papadimitriou, C. Papazoglou, A.S. Pomportsis. *J. Light wave Technol.* 21(2003)384.

[60] B. Wu, A comparison of optical switches and their applications, in: NFOEC Tech. *Proc., Baltimore, MD, USA,* 8–12(2001)255.

[61] D. Chaires, All-optical switching, in: *NFOEC Tech. Proc., Dallas, TX, USA,* 15–19(2002) 1868.

[62] S.H. Chen, H.M.P. Chen, Y. Geng, S.D. Jacobs, K.L. Marshall, T.N. Blanton, *Adv. Mater.* 15 (2003)1061.

[63] A.M.C. Dawes, L. Illing, S.M. Clark, D.J. Gauthier, *Science.* 308 (2005)672.

[64] A. Erlacher, H. Miller, B. Ullrich, *J. Appl. Phys.* 95(2004)2927.

[65] V. Mizrahi, K. D. DeLong, G. I. Stegeman, M. A. Saifi, M. J. Andrejco, *Opt. Lett.* 14, (1989)1140.

INDEX